YOUR KNOWLEDGE HAS VALUE

- We will publish your bachelor's and
  master's thesis, essays and papers

- Your own eBook and book -
  sold worldwide in all relevant shops

- Earn money with each sale

Upload your text at www.GRIN.com
and publish for free

Kathrin Müller-Rees

UNICUM.de – Die Wissensreihe

UNICUM.de

Band 18

The Influences of Grazers in Biodiversity of Insects

GRIN Verlag

Bibliografische Information der Deutschen Nationalbibliothek:

Die Deutsche Bibliothek verzeichnet diese Publikation in der Deutschen National-
bibliografie; detaillierte bibliografische Daten sind im Internet über http://dnb.d-
nb.de/ abrufbar.

Imprint:

Copyright © 2012 GRIN Verlag GmbH
Druck und Bindung: Books on Demand GmbH, Norderstedt Germany
ISBN: 978-3-656-28913-5

This book at GRIN:

http://www.grin.com/en/e-book/202421/the-influences-of-grazers-in-biodiversity-
of-insects

The Influences of Grazers in Biodiversity of Insects

Fundamental and Applied Biology of Insects

Written by Kathrin Müller-Rees

29.09.2012

Summary

Grassland in Europe and other parts of our world offers many animals an appreciable biotope. Not only insects also spiders adopt this specific vegetation influenced by dry and poor soil as their habitat. In Europe the grassland disappears more and more. Many of these areas were under farmers hands and are left fallow nowadays. After a certain period it will happen that bigger herbals such as trees and bushes obtain the fallow area.

In this essay a certain study case made from the Swedish University of Agricultural Science will show how grazers influence the diversity of insects. Grazers such as cattle, sheep and horses are introduced as mowing animals with the goal to avoid the growth of larger plants as trees and bushes.

Table of Content

1. Introduction

All over the world the nature is featured with different biotopes. High Biodiversity within plant species and animals can be provided in all those variation amount of biotopes. European grassland, as one of the category of these areas, offers indeed good conditions for a high diversity of species and their habitat. The good condition for breeding leads particularly birds and invertebrates to live in this area. Over the last year's most European grassland was regulated by grazers or maintained through cutting. Due to agricultural changes in its practice and land use grassland more and more disappeared significantly. Nowadays Europe showed an enormous decline of grassland area which declined 12, 8 %. [6]. Caused by agricultural land using many fields are used intensively and so the physical condition of soil changed completely. [7] Wind erosion influenced more and more the grassland area. Caused by the wind the sheet erosion increased very fast, so the area of grassland is forced to be changed all the time. This fact is highly interested for insects which have to adapt the quality of the biotope in every moment of their life. Not all species are able to move quickly from one place to another if the biotope is on the way to change its collocation. So grassland with its high level living quality is an eminent biotope for all insect species. Flowers, grasses, plants and the so plant covered soil provide nutrition, hiding places and reproduction area. Grassland areas can be characterized by different kinds of their using such as grazed areas or non-grazed areas. The different areas of grassland all over Europe can be also separated in dry locations in middle Europe such as so called primary grasslands. This form of grassland is found naturally by erosion of wind and water. Primary grasslands often are located on rocks and consequently they do not have a high significant under layer or a high amount of soil layer. Secondary grasslands arise as a consequence of human. In the earlier times our ancestors maid their fields and cut trees in forest without controlling the risk of the decline of natural grassland area or even erosion. Cutting trees and farming the land conducted that areas become dry. Wind bowed threw the areas, the soil became sandy. As a consequence of disturbing the natural way of the existing biotope changed into grassland areas. At least the various types of grassland differ mainly in their size: while natural expanded grassland so called primary grassland is mostly found in small covered areas secondary grasslands are characteristic in their wide expansion and hence big areas. Both named vegetation types tend to be as forest if they are unused as grazing area or

3

not influenced anymore by erosion of wind and water. As far as the vegetation type of grassland decreases more and more the number of endangered insect species can be influenced of this fact one day. Especially the diverse group of arthropods reveals enormous variations in their behavior in different seasons, size, their possibility of moving or not, the trophic level or even their strategy of life history. Arthropods also have requirements for habitats of particular successional phases or the vegetation in their different structural characteristics. [1] The effects of influencing grassland areas by agricultural methods are for Arthropods therefore very significant. This paper will treat the management of characteristic grassland focusing on the habit of insects.**Fehler! Verweisquelle konnte nicht gefunden werden.** [4] In general Agricultural and forest management by human plays a big role of their habitat. Different benefit types of natural resources such in forestry and agriculture changed the landscape dramatically. The way of using grassland as meadows for cattle, sheep and horses provoke that the affected vegetation type changed. Shortcutting of grazers induced an adjustment in grass length and density. Different types of grazing such as intensively grazing by cows or less grazing influence have a determined effect on the basic components of insect species. [1] But grassland is not only an area where grazing animals are searching for nutrition. Grazed areas without human influence appear in forest areas or such as isolated biotopes. As a place of high biodiversity in species of animals and insects the characteristic grassland takes an important role in the network of nature. In the concrete example below grazed areas by cattle will be compared with non-grazed areas without cattle with a view to the amount of insects. The hypothesis that species richness is depending on the abundance of plants will be tested in the following disposal. This essay will also point out the requirements of the insect biodiversity of beetles, butterflies and grasshoppers [11] towards the vegetation area in grassland.[8]

2. Description

In the beginning of the description some definition should be clarified. So the term of characteristic grassland is often used. This specific area is mostly covered by grasses as the dominant vegetation. Trees and large shrubs don´t exist in this vegetation type. Climate as an important factor influences this area with a high differential of temperature during

4

wintertime and summertime. Typically it consists of warm and wet summers followed by cold and dry winters with heavy frosts. Species such as many bulbous plants are dominating this vegetation type. Grasslands are often characterized the presence of many wetlands. However offers a high diversity of insects' species and plants a convenient place for living. Due to the specific and characteristic conditions it has a multivariate flora and fauna which will be told in following paragraphs.

Characteristic grasslands are defined as dried areas inhabited by grasses and dry plants which don´t need much water. [6] Such plants and grass species are *Bromus erectus* and Tor-grass (*Brachipodium pinnatum*), *Anemone silvestris*, *Euphorbia cyparissias* and *Trifolium pratense* (et cetera). Trees species are rather rare in this area because of the soil condition. Most times trees are not able to grow because of grazers. Sandy soil causes also a drought area and the water abundance is decreasing. At least strong weather conditions such as in summer or winter with gelid winters complicate most plant species to grow in this vegetation area. Different grass species are in this case the dominant species because they are able to adapt to the ruff conditions in a strong way. [7] Flowers also exist in grassland. As already mentioned Trifolium pretense and Annemone silvestris settle also down in this vegetation type. Last mentioned plants are very important for insects species like honey bee. Old grassland used as active agricultural field by human has a very good physical condition. Field capacity is on a good level and the relationship of water amount in this area also is in a good shape. Due to the good network of grass roots in the soil old grasslands are also highly protected against wind and water erosion. [6] The soil of grassland mostly tends to be calciferous. A feature of calciferous soil is the high amount of calcium carbonate. The percentage of lime can be up to 40 percent. This condition causes a high level of species richness in plants because the plants need a balanced PH-value in the range of five up to seven point five. This volume is very important for their gathering of nutrition. Calciferous soils are soils such as black cotton soil, mixed rendzina in Aeolian silt deposit areas, rendzina on lime stone, arid sol and soils on clay, calciferous brick earth or also glacial drift.

Different experiments in different areas showed that number of specific orders of insects always occur in the same way.

Many hymenoptera were found in these plots. Apoida were mostly found in areas of grassland with a high amount of flowers. Anthophila such as honeybees are generalists and

5

search for their progeny nectar and pollen on flowers. Therefore honeybees were often seen in heigh level grassland with flowers. Also their family member the so called specoid wasps (such as ammophila sabulosa) have their habitat in grassland area. Both species have an interesting method of getting aliments for themselves and their progeny. In this connection Westrich found out that anthophila only feed on nectar of the same plant where they feed on pollen before. Even so their larvae are dependent on herbal nutrition such as pollen. [3] Beetles (Coleoptera) are also mostly presented in areas of characteristic grassland. Due to their high diversity of species and their good sense for finding niches in the ecosystem they fill in the grassland with their variety. Their mobility also allows beetles to settle down quickly on a new location. The beetles´ population rate is very high due to the named circumstances. Hence beetles have a high dispreading rate. Beetle species also have a special connection to their host plants as far as they don´t move that much around such as honeybees do.[11] Most observations of different experimental plots showed species of weevils (Curculionidae), seed weevils (Aponidae) and leaf beetles (Chrysomelidae).**Fehler! Verweisquelle konnte nicht gefunden werden.** Some beetles are herbivores in their larval stadium. During this time beetles are mostly living from plants such as grasses and flowers. One third of adult beetles also are herbivores such as eating pollen on shallow flowers [3] and other plants.

The adults of hoverflies as a member of the family of so called Diptera also feed on pollen. So there are also correlated to grasslands with a high level on herbal nutrition such as flowers. But not only nutrition is important for a habitat on grassland area. Hoverflies also use this area as a place for reproduction and hiding place. Larvas of hoverflies consume aphids [3] and therefore they are predators in contrast to the adult hoverflies.

Butterflies (Lepidoptera) mainly consume nectar of flowers. This is the reason why they are often related to grasslands which have a high amount of flowers. Even the larvae of butterflies have a high quantity on feeding of herbal nutrition due to the fact being herbivore. [10]

Different species of grasshoppers (Orthoptera) are also located in areas of characteristic grassland. As well as beetles they are able to move over large distances due to their big legs.

Spiders (Arachnida) use a web to get their prey caught. Hence there is a necessity to find places and build webs for hunting little insects catching them in their webs and kill them with poison. Due to the facts that some little insects tend to live in flower covered areas their predators such as Arachnida follow them consequently and try to hunt them.[9]

The spread of insects in different areas depends on the species. Mainly grassland area appears in different sizes. So different insect species found various methods to move such as the moving grasshoppers do. Insects with small size are often not able move over large distances and only can handle to move from one place to another in a moderate way. Single insects are sometimes able to move away from one area. Moving the whole population of one species tend in contrast very difficult. This also costs a lot of energy for the animals and not every single insect would arrive at the new area.[10] Populations are normally nothing more than a collection of individuals but even then the group of insects stays at the common place and does not risk a high drop of population when moving to another place. The quality of grassland is also an important factor for the population rate of insect species. Grassland offers nutrition and place for reproduction.

But due to the fact that characteristic grassland more and more decreases also the bandwidth of living area decreases. Insects do not find their spot for living and so the . population of them decreases.

Herbivores need the grass because it is the only nutrition for them. It provides energy and it is therefore the motor of life.

3. Method of the concrete example

Some of natural characteristic grasslands are managed by humans. They use that area as grazing and mowing place for agriculture. Those areas could be well seen in regions such as in Eastern Europe. The so called Poloninas located in the Ukraine provide insects a large biotope and it is used by humans in the same times. Sheepherders tend their flock of sheep and goats on that grassland vegetation.

Now, in the following case [11] will be shown how the grazers and the way of mowing and burning of grasslands have an influence on insects.

The department of Ecology of the Swedish University of Agricultural Sciences performed a study in eight different areas in their country. All areas were covered in a same amount with forests, characteristic grassland, arable land and water covered quarters. In those areas which are randomly distributed three types of management plans were followed. Grazer like cattle divided the areas into acute grazed area and less acute grazed area. In the end there existed grassland with many and grassland with few ruminant animals (such as cattle). Cattles which were kept on intensively managed pasture were managed in agri-environmental schemes hold with management contracts and compensation in an economical way. Economical way means that cattle had to keep on the grassland due to a specific management plan of farmers.

Each experimental plot was fallow for the past ten years. All experimental plots were chosen far enough from each other in a distance of at least ten kilometers due to the fact of the important role of independence value. However those experimental plots were situated close enough to get the same pool of to be determined species. The amount of insects in each area was measured for four times during summer. The three grazing intensity regimes (acute, less acute and non-acute) in each of the eight areas were visited on the equal day and randomly observed. The time of observation was between 7th of June 2004 and 20th of August 2004.

For this study following groups of insects were observed. Distinguished only in their history of origin those insects may also live in other habitat or management. True butterflies, Rhopalocera, Zygaenidae and burnet moths as butterflies were included in the survey as well as flower-visiting bees called Apidae, hoverflies (Syriphidae) and last but not least the order with beetles as species (Coleptera).

Vegetation height, the cover of litter, microstructures, the differences of vertical temperature and also abundance of flower and the number of flowering plants were all measured as indicators of management intensity such as characteristics of vegetation. Grazing intensity was influenced by vegetation height and the accumulation of litter. Due to this fact the vegetation height played therefore in the scheme of grazing intensity as a continuous variable an indicating role. Different site-specific characteristics were also estimated such as the covered area of tree and bush, and the size of pasture. This was done because trees and bushes were indicated to shade the area of the grassland and so therefore influences the study case. Also the structure of the ground was measured. As seen on the indicator plants such as Asteraceae, Fabaceae, Plantaginaceae and Dipsacaceae the measurement of soil yields soils of sandy-covered and stony quality.

In the end of the observation many insects groups were found. Bees (Apoidae), hoverflies (Syriphidae) butterflies (Rhopalocera, Zygaenidae) and beetles (Coleoptera) were observed being in flowers and in their biotope of characteristic grassland.

The numbers of insects were measured in those eight different areas and so the specific characteristics such as pasture size, the composition of the ground and the percentage of trees and bush covering had a significant role in this experiment. The experiment was influenced by the influence of different eutrophic soils. This fact leads that the plant vegetation differs in the different areas. Besides the eutrophic influence there were two measures of ground structure in sand-cover and soil with stones measured. This kind of recording reflected also the human activity concerning experimental plots besides streets and road lengths.

As a general result of this observation there were 3613 insects found in relation to 9467 visits. Those insects covered six orders, 54 families and in the end even 294 species. The observation number per order depending of different species was different. The experimental plots had 35 species of Coleoptera, 92 species of Diptera, 212 species of Hymenoptera, 29 species of Lepidoptera, 16 species of Heteroptera and one species of Neuroptera.

The result on managed areas with grazing and mowing showed that the vegetation of plants and animals was less littered in the intensive grazing area versus grazed areas with low-intensive grazing. The mean of vegetation height of those categories was lower and there were less litter compared to abandoned grassland. The management regime did not influence the temperature, plant species richness, flower abundance or even microstructures. However insect groups, abundance and the richness of species were observed to be correlated in a significant way. The management categories did not force butterflies and bees to be abundant in a significant way or it did not influence their richness of species. So far richness of hoverfly species and its abundance were on its highest level in low intense and disused grasslands. Latter named area was also in favor of biotope for beetle species.

The number of the variance in these eight areas was put in a triplot diagram. So the amount of variance in species composition of different insect groups is set in relation to different environmental variables at three grazing intensity levels. The plots are measured in different

9

spatial levels: local sides and landscape were different important factors. The results of the diagram were shown in a scale with a highest. At first bees were counted in different observation plots. Different special levels in this area were put in a trip lot. It shows that apoidea have a high variation of 13 percent when flower abundance shows up. On grassland the total variance got up just until 10 percent. Watered areas showed the lowest variance of bees in the landscape area. For bees we conclude that the area where bees find their best habitat is covered with flowers. This is related to their nutrition need of pollen and nectar. The order of Lepidoptera prefers an area with an elevated vegetation height. The variance amount is here 15 percent high. However the order of coleoptera with all beetle species is more discovered in areas with a lower vegetation height. The variance of beetle species is therefore only seven percent. This is probably caused by the habit of beetles. With their sedate body they are not able to move quickly from one place to another. Additionally to this fact it becomes even more difficult to move of the grassland is saturated with tall grown grasses. Syrphidae occur also mostly vegetation area with a significant height of grass. They also prefer this area such as Lepidoptera because high grassland offers good hiding places and much nutrition on flowers with nectar and pollen. The amount of variance of Syrphidae is at a level of twelve percent on a high vegetation height. Syrphidae also take place next to roads. See Figure 2 in [10]. The response variables in this concrete case species richness and abundance of plants and arthropods are depending on each other, so the Hypothesis on the depending level of species richness and plant abundance was right.

Due to the decreasing grassland areas over Europe there is no large expansion of insects possible anymore. But decreasing fields' sizes will advance animals such as spiders to live there and also phytophagous beetles adapt on this characteristic grassland. [9]

4. Discussion

Looking to the concrete experiment in Sweden this is one method how to get to know the diversity numbers of species by separating the experimental plots in eight different areas. Even grazers were separated in two groups. Ones cattle grazed and on the other hand horses were used as being grazers in this area. But like all observations it should be pronounced that the experiment usually depends on so many factors that a statistical test is mostly counterfeit by different numbers and the calculated solution does not correlate to the natural property.

Management of landscape on different intensively grazed areas had a significant influence on the diversity of insects and spiders, but also on plants. [10]

Higher diversity of insect in extensively grazed than in intensively grazed meadows. Extensively grazed were meadows with 1.4 cattle per hectare and intensively grazed 5.5 cattle per hectare. The difference between the diversity of spiders and carabid diversity was not significantly in very extensively grazed (0.5 cattle per hectare) and extensively grazed (1.0. cattle per hectare). Tscharntke´s observation took place in Hungary. Several studies concluded that there is a positive effect on diversity of arthropod species. [3] With a view to the outcome grazed areas do not influence in a significant way the spread of the richness of different insect species.

The interactions of insects with grasses are very strong. All insect species living in grassland area adapt grasses as living area, nutrition and a local place where reproduction can be guaranteed. Grassland provides also in contrast to a tree settled area a high level of open landscape without any prevention of mega herbivores. Those are able to pretend the insects from feeding themselves or even getting eaten by the mega herbivores. Mega herbivores are part of the agricultural use by human and also as wild living animals such as deer in forests. So on one hand arthropods tend to be in a higher amount when there is no influence by grazers. This influence could be in an intensive or extensively way.

On the other hand those grazers are necessary to keep the growing and invading tree and higher plant species away from this biotope.

From an ecologist´ point of view the nature should be kept as itself. Agriculture covers more than half of the terrestrial surface of the Earth. Those areas are excluded for an effective long-term conservation. Agricultural used landscape can be seen as a desert on the biological point of view, or as an area which is not valuable for insects. [4] Nature should regulate and balance itself as good as possible and humans should give the area which was occupied in earlier times back to nature. Many insects are more and more abandoned because of loss in biotopes. Therefore the European Union created the Red List of endangered species which also lead characteristic grassland to be named as a protected reserve. Grassland areas are important for insects because it provides an area where also

11

endangered plant species are able to grow. That again let insect species come to live and reproduce themselves in this area. The population of insects depends on the opportunity of living area and so therefore their reproduction and feeding necessity is influenced by the environment of living.

Some endangered species such as grasshoppers are living in the grassland areas. It can be assumed that threatened species prefer to live in this area where rich nutrition and a high biodiversity of feeding such as flowers and other plants are living on this biotope. An overgrazing in this area would also influence the abundance of certain no threatened and threatened plant species. The livestock increases and on the same time the mobility of the livestock decreases and tends to be low. Plant diversity and biodiversity of insects decreases too and grassland degradation can happen.

From the point of view of farmers as economists would like to make profit of these areas. They support the cattle feeding on characteristic grassland what not always means that there will be an abundance of insects. As a result of these observations and observation of other contents in literature you can conclude that the only species which takes a benefit of mowing is the family of grasshoppers. [3] It even benefits from the continuously short kept grass length because of the ability of a better shifting between two spots. But agro ecosystems also could offer benefits on the economical way and is responsible for important services of ecosystems. Hereby the ecosystem services should be defined as pollination and biological control. [4]

5. Conclusion

This Essay shows that there is a high level of need to keep grasslands all over the world. Insect species such as endangered species of grasshoppers only survive if grassland areas will be preserved from disappearing.

A conflict of agricultural use and natural way of living for insects is not extraordinary provoked. Mega herbivores such as grazers could influence the population rate of herb feeding insects but not in a significant way. Both animal groups also are able to live besides each other. Cattle and horses help to keep the grassland area open and protect it against invading plant species which are able to cover the area which is indispensable for life of insects.

 Decreasing fields' sizes caused of different factors in the last years will however advance animals such as spiders to live there and also phytophagous beetles adapt on this characteristic grassland. [9]

As a conclusion it is important to say that protection of biodiversity needs a perspective in landscape and should combine nature conservation and farming on a balanced way. Nowadays many projects such as "biospheres" are introduced with a perspective to follow the goal of a balanced way between the economical and the ecological perspective.

6. Sources

[1] A.Hudewenz, A.-M.Klein, C.Scherber, L.Stanke, T.Tscharntke, A.Vogel, et al. (2012). Herbivore and pollinator responses to grassland management intensity along experimental changes in plant species richness. *Biological Conservation.*

[2] A.Körösi, P.Batary, A.Orosz, D.Redei, & A. (2012). Effects of grazing, vegetation structure and landscape complexity on grassland leafhoppers (Hemiptera: Auchenorrhyncha) and true bugs (Hemiptera: Heteroptera) in Hungary. *Insect Conservation and Diversity.*

[3] B.A.Woodcock, J.M.Bullock, S.R.Mortimer, & R.F.Pywell. (2012). Limiting factors in the restoration of UK grassland beetle assemblages. *Biological Conservation.*

[4] Batary, Holzschuh, Orci, Samu, & Tscharntke. (2012). Responses of plant, insect and spider biodiversity to local and landscape scale management intensity in cereal crops and grasslands. *Agriculture Ecosystems & environment*

[5] Collinge, P. O. (2003). *Effects of Local Habitat Characteristics and Landscape context on grassland Butterfly Diversity.*

[6] J.S.Pryke, & M.J.Samways. (2012). Ecological networks act as extensions of protected areas for arthropod biodoversity conservation. *Journal of Applied Ecology.*

[7] J. Low. (2006). The effect of cultivation on the structure and other physical characteristics of grassland and arable soils. *Journal of Soil Science*

[8] P.Batary, A.Baldi, D.Kleijn, & T.Tscharntke. (2011). Landscape-moderated biodiversity effects of agri-environmental management: a meta-analyis. *Proceeding of the Royal Society B- Biological Sciences.*

[9] P.Dennis, M. I. (1988). Distribution and abundance of small insects and arachnids in raltion to structural heterogeneity of grazed, indigenous grasslands. *Ecological Entomology.*

[10]Pryke, & Samways. (2012). Ecological networks atc as extensions of protected areas for anthropod biodiversity conservation. *Journal of Applied Ecology*

[11]Skjödin, B. E. (2008). The influence of grazing intensity and landscape composition on the diversity and abundance of flower-visiting insects. *Journal of Applied Ecology*